# Useful Beauty

## TURNING PRACTICAL ITEMS ON A WOOD LATHE

# DICK SING

Text written with and photography by Alison Levie

Schiffer Publishing Ltd

77 Lower Valley Road, Atglen, PA 19310

*Dedication*

*To my grandson*
*Collin Daniel Powers*
*Papa's shadow and Grandma's most*
*expensive toy.*

Printed in China

ISBN: 0-88740-851-6

*Book Design by Audrey L. Whiteside.*

**Library of Congress Cataloging-in-Publication Data**

Sing, Dick.
    Useful beauty: turning practical items on a wood lathe/Dick Sing; text written with and photography by Alison Levie.
        p.   cm.
    ISBN 0-88740-851-6 (paper)
    1. Turning. 2. Lathes.  I. Levie, Alison.  II. Title.
TT201.S53   1995
674'.88--dc20                                           95-22738
                                                           CIP

Published by Schiffer Publishing, Ltd.
77 Lower Valley Road
Atglen, PA 19310
Please write for a free catalog.
This book may be purchased from the publisher.
Please include $2.95 postage.
Try your bookstore first.

We are interested in hearing from authors
with book ideas on related subjects.

# Acknowledgements

The Berea Hardowood Co.
125 Jacqueline Drive
Berea, Ohio 44017
(216) 243-4452
(wood and project supplies)

Choice Woods
9714 Blue Lick Rd
Louisville, Kentucky 40229
(502) 966-3958

Craft Supplies U.S.A.
1287 E 1120 S.
Provo Utah 84601
(801) 373-0917
(woodturning tools and project supplies)

The Hardwood Connection
420 Oak Street
DeKalb, Illinois 60115
(815) 758-6009
(wood and supplies)

One good Turn
3 Regal Street
Murray, Utah 84107
(801) 266-1578
(domestic and exotic woods)

Woodcuts Ltd
7012 Highway 31
Racine, Wisconsin 53402
(414) 681-1986
(hardwood, carving and woodturning supplies)

# Contents

# Introduction

One of the great joys for the woodturner is the opportunity to create an object which expresses the beauty of the material, while serving a useful purpose. The transformation of the ordinary into the exceptional makes us practical artists, participants in spreading creative beauty to unexpected nooks and crannies.

The projects I have chosen to share in this book are all small treasures, rather than large, dramatic ventures. Their charm emerges unobtrusively, sneaking into our awareness without fanfare, and yet once we notice them, we have to keep looking again and again. They do their chores so gracefully, with such style, that we cannot help but be impressed.

Consider the jar lids and bottle stoppers, for example. What could be more mundane than to seal a jar or wine bottle. But a simple turning project, requiring minimal materials, can transform a utilitarian lid into a thing of beauty. The decorated back of a small mirror can be as attractive as a peek into the glass, and no plastic rattle can compete with a piece of apple or pear wood, turned and sanded to smooth perfection. A spinning top, one of the simplest of toys, requires no battery or sound effects to entertain children of all ages.

Try captivating people with the captive rings on the rattle. With care, the rings are not difficult to create, but I find that people are amazed when they realize that it all comes from one piece of wood. I really enjoy watching people try to figure out how it was done.

The weed vases, bottle stoppers, and jar tops provide a chance to experiment with designs. I have shown some pleasing examples, but don't be afraid to create your own. Look around you for shapes that would be fun to try. Study pictures, or sketch your own ideas. If while you are turning, you make a mistake, stand back for a minute to see what new shape is possible. Let the wood and your imagination speak to you. There is no right answer in this class. Just a lot of fun and satisfaction.

All of these projects make wonderful presents and easy sellers at craft fairs. They are something a little different, a little unusual, and very useful.

# Bottle Stopper

Turning a bottle stopper is an enjoyable, functional project which is simple to do but provides design challenges. The stoppers are extremely popular and make great gifts. While the actual execution is simple, by using different woods, materials or shapes, tremendous variation is possible. Work to blend these three qualities to create a pleasing effect.

When you do a number of these bottle stoppers, you become very conscious of shape. The project also teaches you to work lightly with your tools.

Because you are working on a dowel which is much more sensitive to chattering and breaking, you become much more conscious of proper tool control. While there are other ways to hold the stopper while turning, the method shown here works well for turners with minimal equipment and experience. I myself use this same technique.

I am using cocobolo for this stopper because this colorful wood is fine grained, taking a nice finish, and has a striking variation of grain, all of which make for an attractive project.

This project requires a block 1 1/4" to 1 1/2" square, by 1 3/4" to 2" long; a 3/8 by 2" dowel, and a cork.

With the awl mark the center so that it will be easy to find when drilling.

Find the center by using a straight edge to draw a line from corner to corner with a pencil. Repeat using the other two corners.

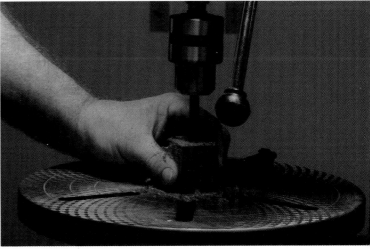

Drill a hole approximately 1/2 " deep with the 3/8" drill bit.

Apply glue to the hole and the end of the dowel and seat the dowel using a mallet or hammer. Allow it to cure.

Put a drill chuck in the headstock, and put the dowel for the stopper in the jaws leaving enough space between the stopper and the jaws to face off the bottom with a parting tool.

You don't need to take much off, just enough to clean up the surface and make it flat so that the cork will fit cleanly on the bottom of the stopper.

Loosen the chuck and seat the block tightly against the jaws. Firmly hand tighten. Now bring up the tailstock using a 60°

center. With the lathe running and the tailstock loose, make contact, allowing the point to seek its center and tighten the tailstock. The tailstock provides support eliminate chattering and help prevent the dowel from breaking. Do not apply too much pressure or you will drive the point in too deeply and damage the block.

Using a gouge, clean up the block     until it is fully round without any rough surfaces.

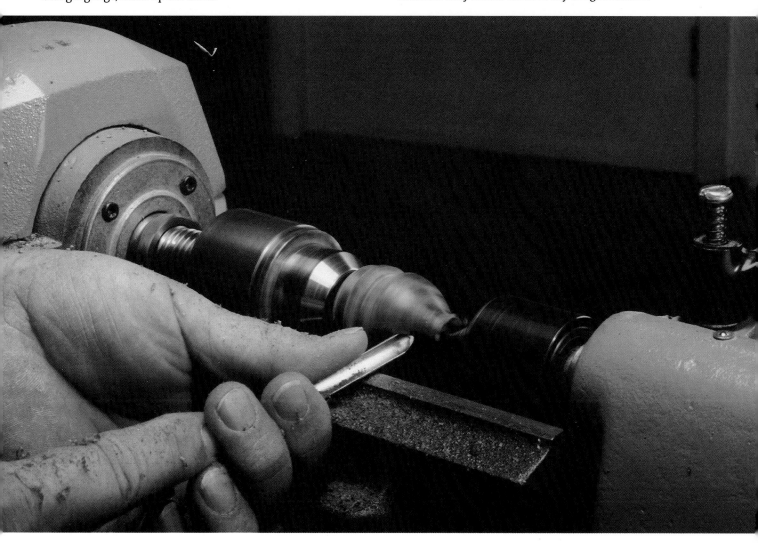

Using a 3/8" spindle gouge, rough out the shape of the stopper, cutting from the large diameter to the small, in effect, cutting down hill. Use your imagination and creativity to create the shape. To choose a shape, you can develop the shape while you are turning, or you can sketch a shape in advance to use as a guide while turning. Photographs or design books can give you ideas for shapes.

If you make a mistake, don't give up. Stand back a little and see what new shape is emerging and work with that. That's how new designs evolve.

Continue to refine the shape. This shape can be difficult, so if you are new to turning, you may want to start with something easier.

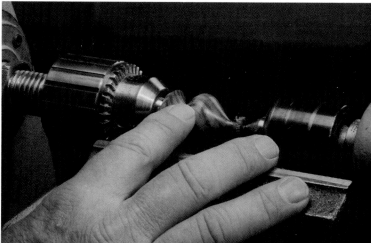

Stop the lathe and check to make sure that your piece is smooth with no torn grain or chipped edges. Now is the time to fix them if you find anything wrong.

One of my favorite shapes is the minaret or bishops cap which I am making here.

Cut off the end, using the tip of the gouge, working from both directions.

Be careful not to tear the grain in the end of the stopper. To do this use light cuts and pressure. Torn grain would leave deep pock marks which are difficult to sand out for a perfect finish.

Make sure all the imperfections are out with the initial sandpaper before proceeding to the next grit. I stack them in my hand from coarsest to finest so I do not have to check which one to use next. I can proceed from one to the other without stopping the lathe. Sand up to the shoulders using the edges so as not to destroy the crispness of your design.

Progress.

In order to remove any concentric rings, I stop the lathe and sand lightly with the 400 grit in the direction of the grain.

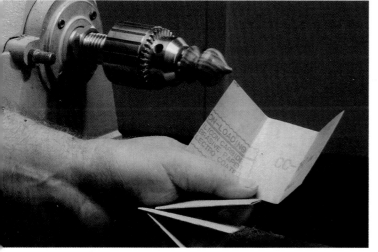

To sand, I cut my sheets into sixths and fold the sixths into thirds, creating a convenient size with a lot of sanding edges. I start with 180 grit, progress to 220, then 320, and finish with 400. If your surface is a little rough, you may want to start with a courser grain before using the 180.

Loosen the chuck lightly and pull out the stopper enough so we can finish the base, and retighten the chuck by hand.

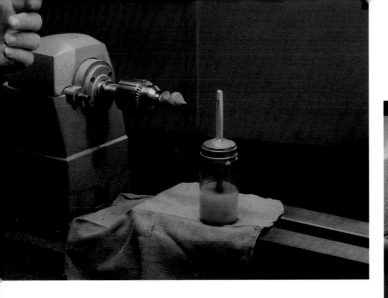

With a cloth, cover the lathe bed to protect it, ( I like clean tools)

Apply a liberal coat of Deft™ satin wood finish.

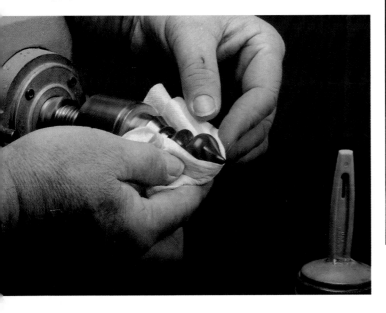

With the lathe stopped, use a good quality paper towel to wipe off the excess Deft™.

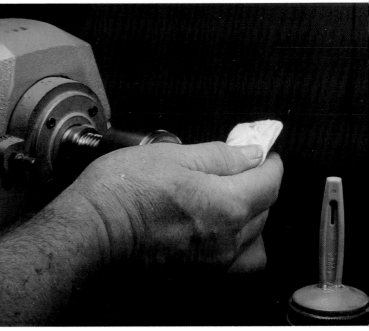

After covering the piece with a paper towel, turn on the lathe and buff as you would with French Polish. Be sure you cover the piece before you turn on the lathe so that you do not spray the Deft around. If you are not satisfied with the finish, apply another coat and buff again.

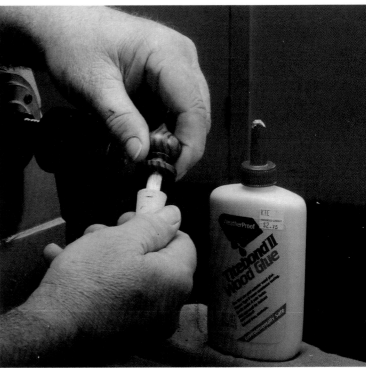

Put glue on the dowel and twist the cork on until it seats on the stopper.

Using a bandsaw, cut the end of the dowel off so that the dowel and the cork are flush.

To finish off the end of the cork, I place the stopper in the jig, square up by eye to a sanding disc,

and sand the end of the cork and dowel.

I use a jig that I made to hold the tapered cork square to a bandsaw or disc sander. The jig is made from a piece of wood approximately 1 inch thick by drilling a 7/8" hole 1/2" deep and a 3/4" hole all the way through on the same center line. This allows you to hold the base of the stopper square to the jig so the bottom of the cork can be sawn and sanded evenly.

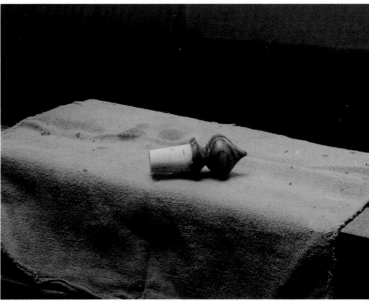

Its a regular cork with a very special top.

# Stopper With A Captive Ring

This project is a little more difficult, but is a great novelty item which can show off your skills. It is fun to see the look in people's eyes when they realize that it is all made from one piece of wood.

I chose Gancalo Alves for the wood because it is fine grained, attractive and takes a good finish.

Making captive rings requires a fine grained wood that does not split easily.

To make a stopper with a captive ring, you start with all of the same steps used to start making the first stopper.

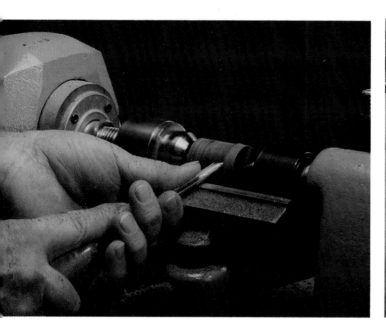

Rounding the block.

Leave a wide bead to create the ring later in the project.

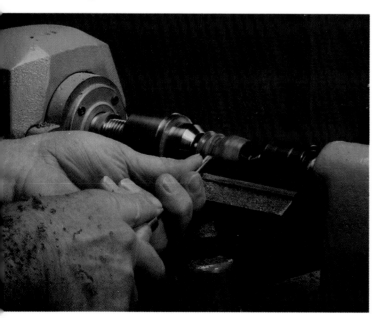

Roughing in the shape and positioning the ring.

The easiest way to make captive rings is with a set of captive ring tools which consist of a beading tool (center), and a left and a right ring tool. I will be demonstrating the method using these tools.

An alternative is to form the ring with a gouge and release it with a reworked dental pick or homemade tool.

Using a small parting tool, strike the width of the bead first on one side then the other.

The tool has been ground flat with the very front edge ground with the relief as a scraper. A left tool and a right tool are necessary.

You need the depth of the cut to be the same diameter on each side of the bead to help guide you as to where to start the cut for the underside of the bead. This will make a more uniform ring.

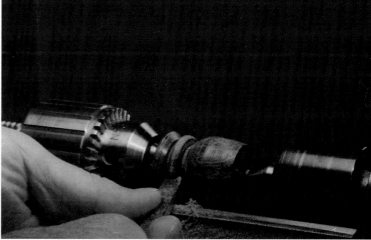

With the beading tool, form the beginning of the bead with a scraping action making sure the tool stays slightly below the center line. If a catch occurs, having the tool below the center line will throw it away from the work. Above the center line the tool will dig into the work.

Using the right hand ring tool

start a cut where the bead and the small diameter meet.

Repeat the same cut on the other side with the left-hand ring tool.

Cut carefully on both sides

so that you do not go all the way through because you want to be able to sand the ring while it is still attached.

Sand using progressively finer sandpaper. Use the edge of the sandpaper to get in under the ring as well as possible. Stop the lathe and check to make sure you are satisfied. This is your last chance for the ring. It cannot be easily sanded after it is cut free.

keep working it back and forth until the cuts match and the ring is free.

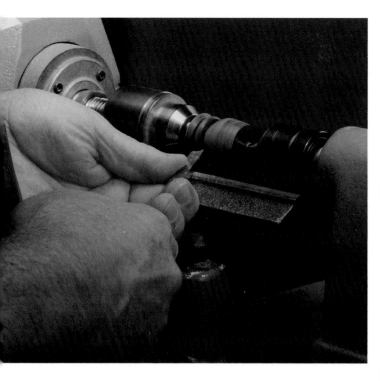

Take the right hand tool and continue undercutting the ring.

Take the small parting tool and moving the ring to the side, make some room underneath it.

Repeat using the left hand tool on the left side.

This should be done on both sides.

With a 1/4" spindle gouge, start shaping this area around the ring.

which is pleasing to the eye.

Switch to a small scraper so that

you can form a shape under the ring

In order to shape the rest of the stopper, you could continue to use the scraper, but you can move more wood

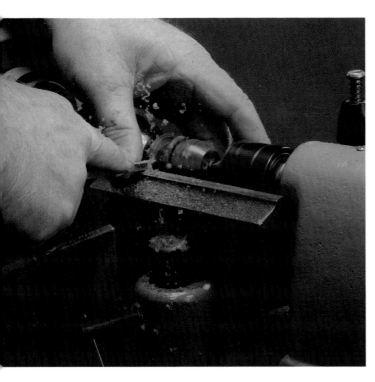

and get a cleaner cut with the gouge.

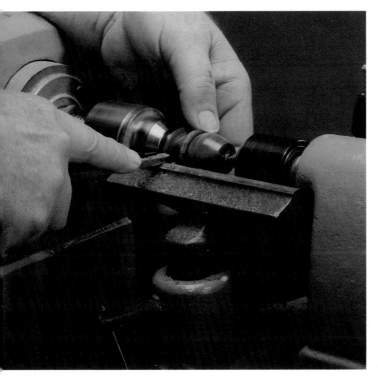

Shaping the body

Finishing the top.

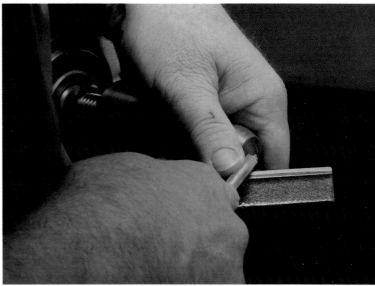

Pull the tailstock away and cut off the nubbin left where the live center was resting. Support the piece with slight pressure from your hand to help control vibrations and prevent chattering.

Sand with the lathe running. Keep moving the paper so that you do not get rings.

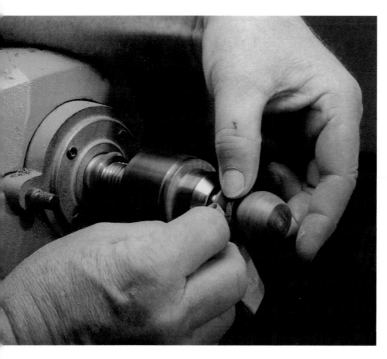

do the ring by hand with the lathe stopped.

When sanding near the ring, hold the ring in mid-air to avoid marking either the ring or the body.

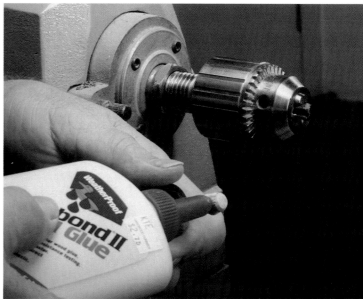

Apply glue to the cork

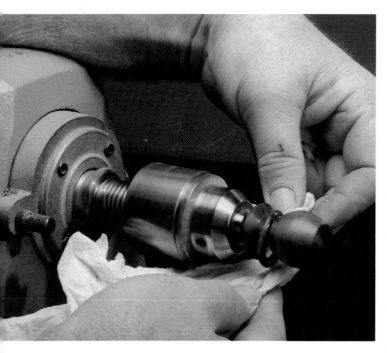

When you have finished sanding, turn the lathe off and pull the stopper away from the jaws of the drill chuck to apply Deft™ to the stopper, including the base. Polish with the lathe running, but

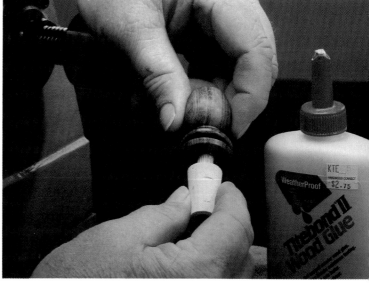

and seat the cork in the stopper. Finish off the same way we finished the regular stopper.

# Golf Ball Stopper

Use a block of wood approximately 1 and 1/4"
square by 3/4" thick. We are using a piece of
cocobolo.

Find the center of the block, first with pencil,

Drill through with the 3/8" drill.

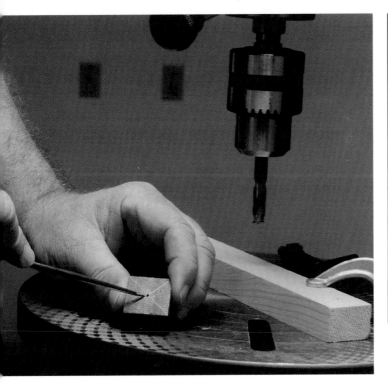

then use an awl to mark the center.

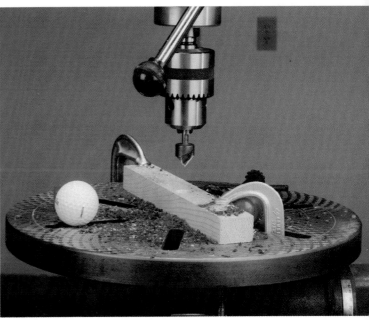

Finding and drilling the center of the round golf ball can be
difficult. Use a piece of scrap lumber and clamp it to the drill
press table, centering it beneath the chuck. Using a counter-
sink, drill a depression into the scrap lumber.

This depression will locate the golf ball perfectly beneath the chuck.

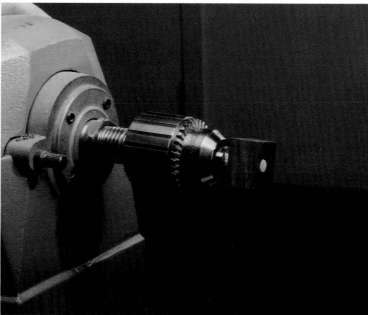

Place a 3/8" dowel in the drill chuck, hand tighten and force the blank onto the dowel. Recognize that only the friction of the blank on the dowel will be sufficient to drive the work piece. You must use a light pressure and clean cuts since this friction does not provide the amount of drive that a solid mounting would.

Place the ball in the depression, lining up any writing, logos, etc. upside down so that it will be right-side-up when the project is finished. Using a 3/8" drill, drill a hole approximately 1/2" deep in the ball.

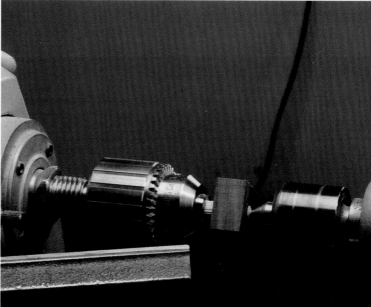

Bring up the tailstock with the lathe running, letting the point find its center. Lock down the tailstock and apply light pressure.

Glue a 3/8" by 2 1/2" dowel into the ball, and wipe off any excess glue so that it does not interfere with the seating of the turned piece later.

With a 1/4" gouge

Reduce the square to a round blank, working from the center to the edges so you do not break off the edges.

start to rough out the form,

Face off the back side to create a smooth surface for the cork.

refining it with the gouge as you progress.

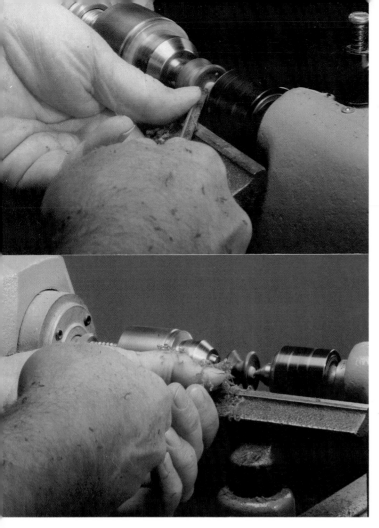

Pull back the tailstock to give yourself room to work and, with a 1/4" gouge, hollow out the top to fit the golf ball so the ball will fit inside the hollow and sit flush on the rim.

If you are uncomfortable going into a cove this deep with a gouge, use a scraper to finish the cove, cutting on one side only so that the tool does not stick or bind between the two walls.

Test the size of the hollow by removing the work piece from the dowel and testing it with the golf ball. Continue to shape the hollow until it fits the ball correctly.

Progress.

Force the dowel into the work piece by twisting the dowel into place.

Sand as before, working through the grits of sandpaper.

Apply a small amount of super glue between the dowel and the work piece to make sure that they stay together tightly.

Apply Deft™ and finish.

Apply yellow glue to the other end of the dowel and slide the cork on. Trim as before.

# Weed Pot

This piece can also be called a dried flower vase. Women love them. I like it because it allows me to use up scrap pieces that are too small for bigger projects. Highly figured wood makes a more attractive vase. I am using a piece of spalted maple because it is flashy, and I had it lying around.

Locate the center of the block and drill a 1/4" hole approximately 1" deep. This hole will serve as the pilot hole for the screw chuck.

Clean up the surface of your scrap block and you are ready to **start**.

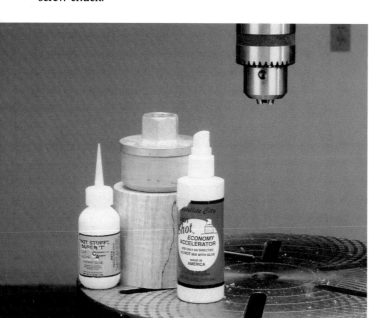

If you are using a face plate, you will need a scrap block which you will screw onto the face plate, and then glue to the work piece. I use hot stuff™ as a glue.

The blank or work piece can be any size or shape. If the piece is erose or irregular, it can be set up on the lathe between centers and the basic shape roughed out. By roughing out the shape between two centers we can move either the drive spur or the live center in the tailstock to shift the workpiece to our advantage.

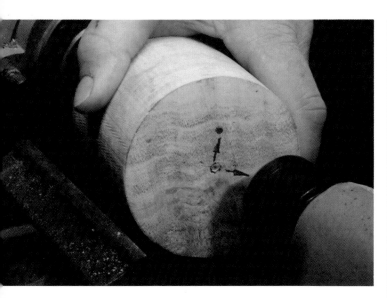

To shift the work piece to balance the piece or to bring out some unique aspect of the wood, you move the center. The marked circle in the middle of this piece represents the natural center. The arrows point to the possible directions to position the blank at the tailstock, although the illustration is exaggerated. Normally small repositioning is all that is necessary. Repositioning the center can also be done at the headstock drive spur.

Reduce the blank to a cylinder using a gouge.

You may need to reduce a lot to get a nice shape. Look at those shavings.

The cylinder.

Checking the shape.

Face off the end with a parting tool to provide a mounting surface for the scrap block.

Reducing some more.

Leave a small tenon at the center

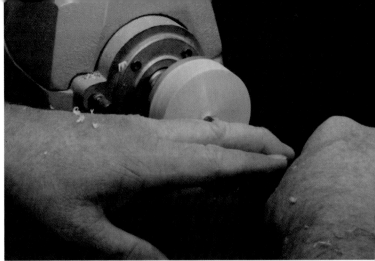

to locate the blank to the scrap block.

Use a parting tool make a hole to match the size of the tenon so that the scrap block and the blank will align concentrically. You can drill this hole if you would prefer, but then you must make sure that your tenon matches a drill bit size.

Use a straight edge to make sure your surface is flat.

Glue the parts together.

Mount the scrap block on the face plate and face off.

Put the face plate back on the headstock. Bring up the tailstock with the lathe running, let it find its center, and tighten.

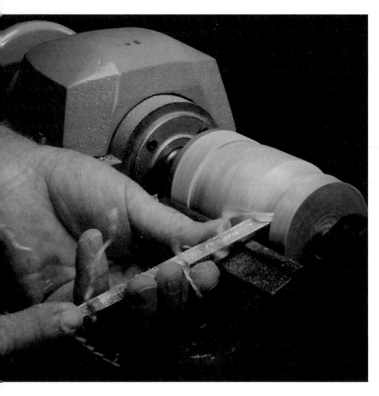

Using a gouge, rough out the shape.

Use whatever gouge you are comfortable with. There is less vibration with the larger gouge, but the smaller gouge can get into places that the larger one will not

Remember to turn from large diameter to the smaller.

Checking the shape.

Using the shear scrape I smooth the shape and get rid of any torn grain or ridges.

A shear scrape is achieved by standing the flute of the gouge at an angle much the same as you would if you were using a pocket knife. You go from small diameter to large, the opposite of a the normal cut.

Some further refining of the shape.

A close-up of the angle used for the shear scrape. The tool is being pulled towards the camera.

Remove the tailstock.

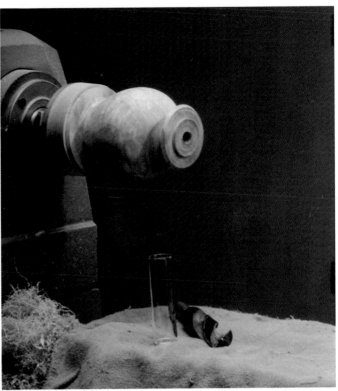

You can insert a glass container into the hole so that the vase can hold fresh flowers in water. I use a test tube which can be found at a pharmaceutical supply house or science store. Drill the hole to fit the size of the tube by providing a slightly larger diameter than that of the tube. The depth of the hole must be slightly shallower than the length of the tube so that the tube can be taken in and out. If you are going to use the tube, the wood must be dry so that it does not warp and bind the tube.

To drill the hole for the flowers, insert the drill chuck in the tailstock with a 3/8" drill bit in the chuck and drill to the appropriate depth. You need a hole which is deep enough to hold the flowers, but you do not want to drill through the bottom of the vase.

Shape the entrance of the hole, creating an attractive rim with a finished look. Try to give the rim a uniform thickness, crisp detail and eye appeal. Use a small gouge with a light touch versus a scraper because you are some distance from the headstock and are unsupported. This condition can lead to chattering.

The mouth of the vase.

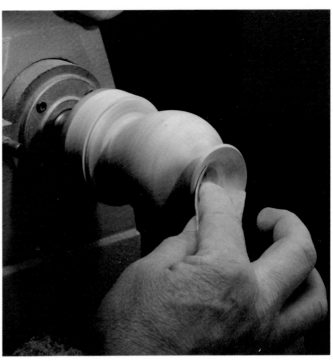

Sand the inside of the neck, starting with the 180 grit sandpaper.

The vase from the side so you can see the depth of the throat.

Finish it with the Deft. Wipe off the excess.

Reduce the block at the foot of the vase

Parted.

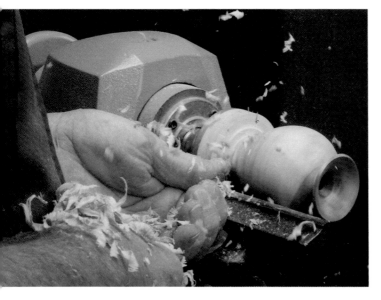

and refine the shape with a gouge.

Put the drill chuck back in the headstock and insert a 3/8"
diameter rod in the jaws. A wooden dowel can be used, but it
must be trued for the vase to run true.

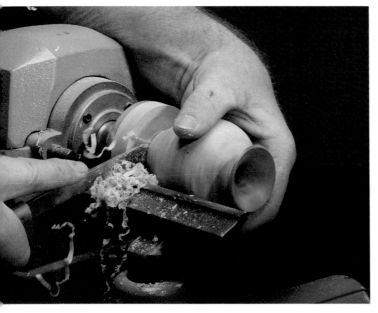

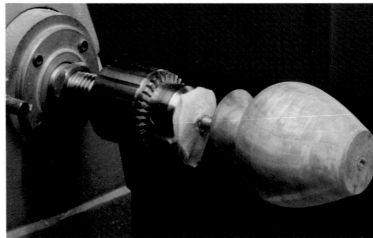

With a parting tool, part the vase from the waste block.

Cut a piece of cloth with a hole in the center to slide over the
rod against the chuck. The cloth will protect the finished
surface of the vase mouth.

Clean the nubbin off the base of the vase with a small gouge and bring the tailstock up lightly for support.

Sand the vase working through the grits. Start with 180 or whatever you need. Make sure that you remove all imperfections with the coarsest sandpaper you are using. The finer grits will only magnify any imperfections. It is always good to sand with the grain with the lathe stopped between grits. The final sanding should always be done with the lathe stopped so as not to create concentric rings.

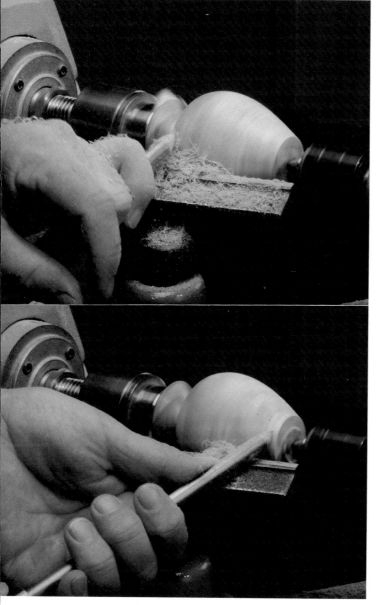

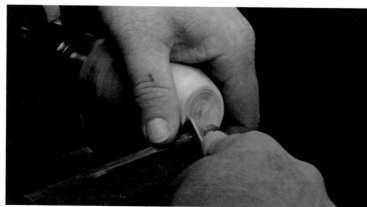

Using the 1/4" gouge, take light cuts to finish the bottom, with a convex curve. The amount of decoration is up to you. Remember that you are supported and driven only by a 3/8" rod.

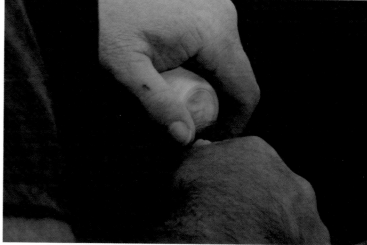

Finish off the body of the vase with the gouge using a shear scrape.

If chattering does occur, lightly hold your hand on the opposite side of the vase to help support it.

Look at how the Deft™ brings out the grain. Since this wood is spalted maple, it normally has punky (or soft) spots which require extra clean cutting. The soft spots will absorb more finish. Keep these spots wet or give them an additional coat or two until they stop absorbing the finish.

With the lathe off, use 0000 steel wool and the tung oil of your choice. (I happen to be using Maloof™ on this project.) Apply a liberal coat of the oil and allow it to penetrate for a few minutes. Then wipe dry. Make sure you let a little oil run into the inside to seal the wood.

Finished. I sign my name on the bottom with a wood burner and identify the type of wood.

# Mirror

I have chosen to use birdseye maple because of its flashy figure, it has a fine grain, is a stable wood and has attractive color tones and contrasts. What-ever wood we use, should be dry so warping does not become a problem.

Attach a scrap block to the face plate and face off so that it is flat.

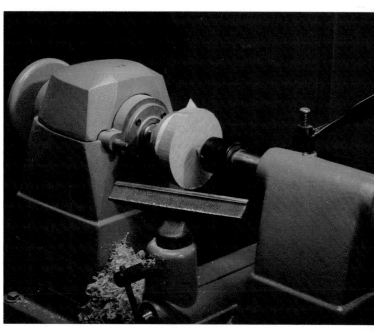

Attach the work piece (blank) to the scrap block and bring the tailstock into position, allowing it to find its center before locking.

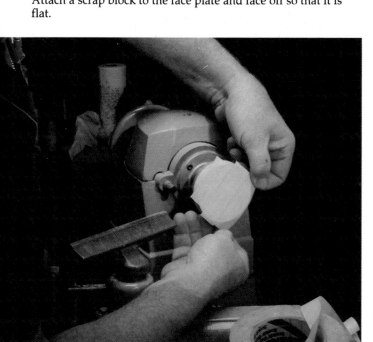

After cleaning both surfaces with a piece of tape to remove all dust, apply double-faced tape to the scrap block.

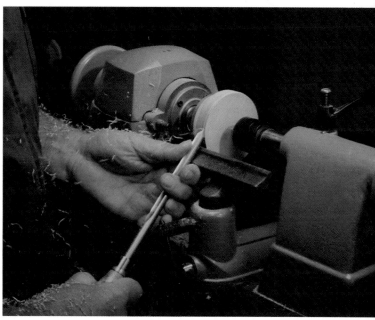

Reduce the diameter and clean up the piece.

Reduce the thickness. I am using a 3" diameter mirror. I have roughed the work piece down to approximately 3 1/2" diameter and 1/2" thick. This will provide room to create an attractive design.

With a gouge, take cuts toward the headstock, thereby applying pressure toward the tape rather than creating a pull on the tape which could separate the work from the scrap block.

Mark the diameter needed for the recess to hold the mirror, but make it slightly smaller than the mirror so you can create a tight fit.

Deepen the hole.

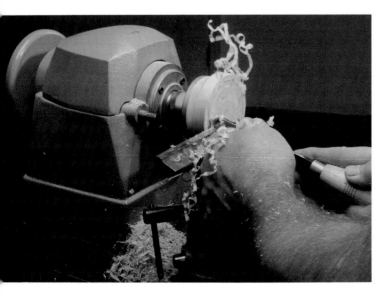

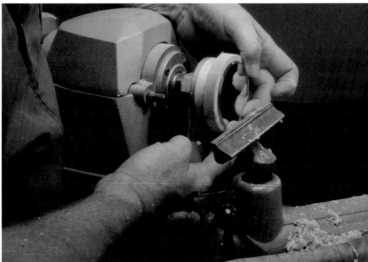

Using a parting tool, start cutting the recess.

Test the mirror for fit. (The recess is not large enough, as we expected since we started with a smaller opening than needed to provide for a correct fit.)

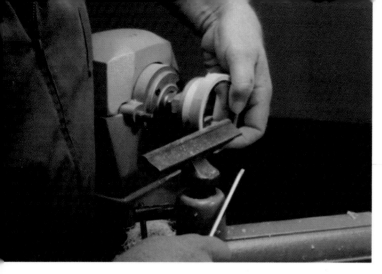

Make the recess a little larger and test again.

Make the recess deep enough so that your final design will be slightly above the surface of the mirror. This will help protect the mirror from damage if it is laid on its face.

Repeat until the mirror fits inside the recess with a small amount of clearance. This is necessary to prevent the mirror from breaking if the piece should warp.

You should now have a flat surface on which to glue the mirror. I am using a small piece of wood with a flat edge that I carry in my coat pocket all the time. This small piece of wood will get into places that a standard straight edge will not fit to show me whether the surface condition meets my needs (flat, concave, etc.).

With a gouge, flatten the bottom, taking light cuts and remembering your piece is held only by tape. The tape is sufficient unless you are heavy handed.

Create the decoration on the face.

Sand the rim. The recess itself is better left unsanded, and the rest of the mirror will be sanded when the mirror is turned over, but this is your last shot at the rim.

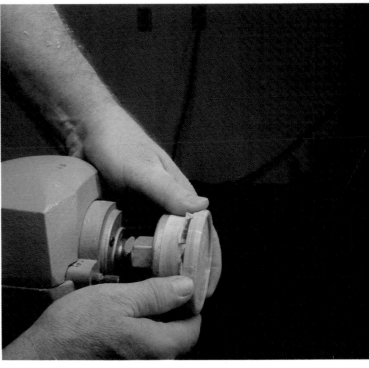

Finish the recess and the rim. The finish on the recess will seal the wood and help prevent warping.

and, with your thumbs, provide constant pressure until the piece pops loose.

Progress.

To remove the piece from the waste block, stand behind the headstock,

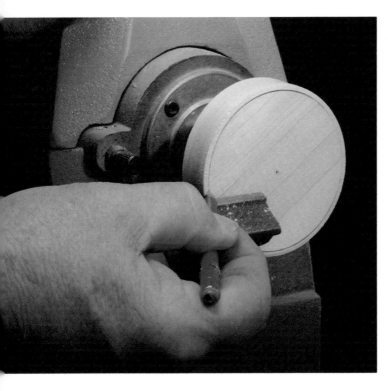

With a piece of scrap on the face plate, pencil a layout line, slightly larger than the diameter of the mirror recess. This is to form a jam chuck which will hold the work piece so we can finish the back side.

Cut to the line conservatively and test fit until the work piece fits snugly onto the jam chuck. This has to be a close fit as it holds the work piece in place while you finish the back side. If you remove too much and the chuck is too loose, try putting a paper towel between the mirror recess and the jam chuck.

Jammed on.

Shape the back.

The back can be left flat or a design can be added by creating beads or lines.

Refining the form.

Sand starting with the 180 grit sandpaper.

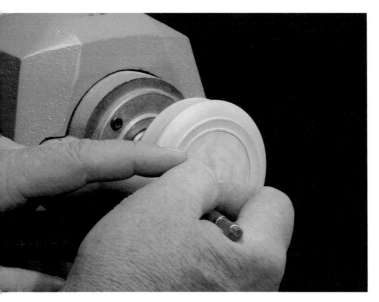

When decorating the back of the piece you may use a pencil to test your design. If you don't like the line there take it off. If you do, make the cut.

Use the edge to sand up to the shoulder without destroying the detail.

Putting in the line using a small pointed scraper.

With the lathe off, apply Deft™ and wipe off any excess.

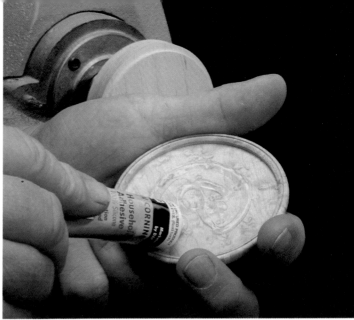

Remove from jam chuck and use oil and 0000 steel wool to obtain the luster of an oiled finish.

Apply silicone adhesive to the recess, but stay away from the edges so the glue does not squeeze out. Silicone adhesive is used because it remains flexible and if there is movement in the wood, it will not break the glass or the bond.

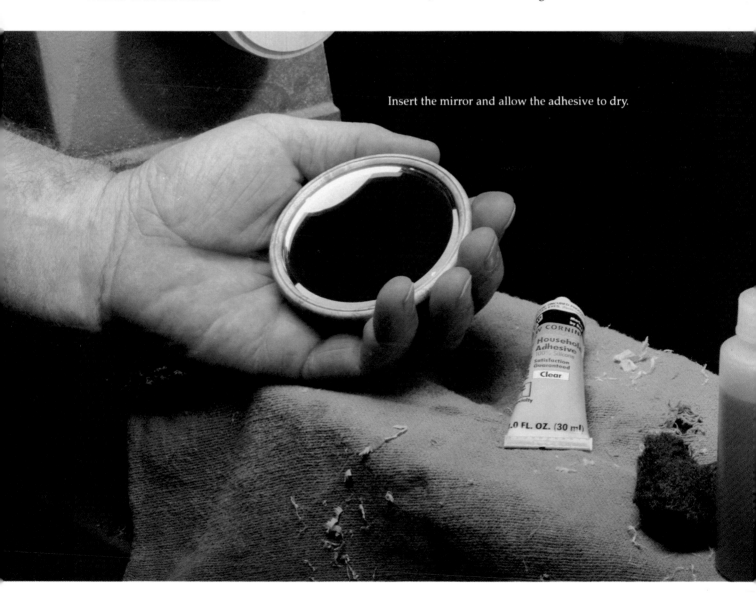

Insert the mirror and allow the adhesive to dry.

# Baby Rattle

The baby rattle makes an excellent gift, as well as an interesting project. This was my grandson's favorite teething toy. I make one for every friend's newborn.

Start with a blank 1 3/8" square by 4 3/4" long. You must use a reasonably hard, close-grained wood so the rings are strong and not prone to breaking or splintering. I would also suggest that you not use an exotic wood as some are toxic and some people are allergic to them. Hard maple and the fruit woods are excellent choices. If the baby is unruly, crab apple would be a fitting choice. I will be using Swiss Pear. Find and mark the center of both ends.

Mount the work piece between the centers and reduce it to a cylinder using the gouge.

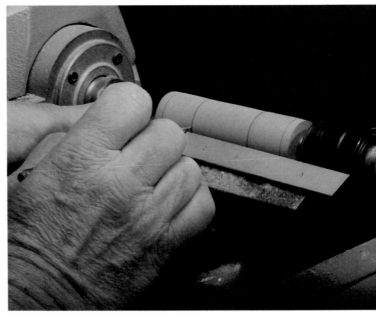

Mark out for a general barbell shape.

Rough out the shape of a ball on each end, creating a barbell shape. Be sure to leave material at the headstock end for the spur center to drive the work piece.

With the beading tool, slightly below center line to avoid digging in, advance the tool forward slowly and form your bead. When finished check to make sure that the surface is clean and tear free.

While creating the basic barbell shape, do not do anything to the center as this area will be laid out for the captive rings. The center section needs to be uniform diameter so that the rings will all be the same diameter.

With the parting tool, go down along the side of each ring.

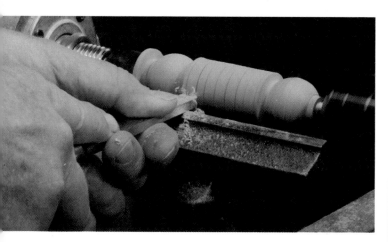

Use a 1/4" beading tool to mark the rings. 1/4" rings are heavy enough that the children will not be able to break them easily.

Keep the indentations the same depth.

Progress.

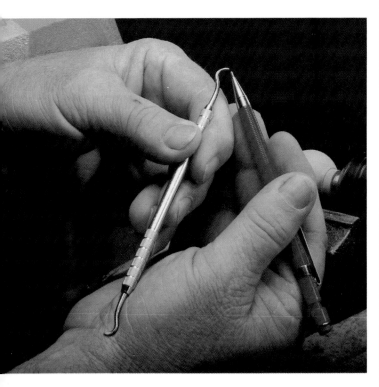

The tool I use to make captive rings is a modified dental pick. It has been heated and wrapped around a 1/4" rod, allowed to cool and ground to its present shape with the very front edge serving as a scraper. One end is ground for a left scraper and the other for a right. The left hand is shown with the pencil pointing at the scraper edge.

Using the corner created by the small diameter and the wall of the ring as a guide start undercutting the ring from both sides, being careful not to cut all the way through.

Sand the rings starting with the coarsest grit and moving progressively through to the 400. Now is the time to sand, because once they are cut free you will not be able to sand them. Do the best you can to get underneath the rings using the edges of the sandpaper.

Working from side

to side, attempt to meet at the middle of the ring which will cut them free, making the bead uniform.

Progress.

Using a parting tool, remove some of the excess wood between the rings. You will now be able to move the rings so you can work the spindle beneath them.

Using a gouge and holding the rings aside, reduce the diameter of the spindle evenly to provide a rattle handle that is heavy enough so that it will not break and proportional to the rattle.

Use a shear cut to shape the balls on the end until they have a nice round surface.

Do this for the ball at each end.

A close-up.

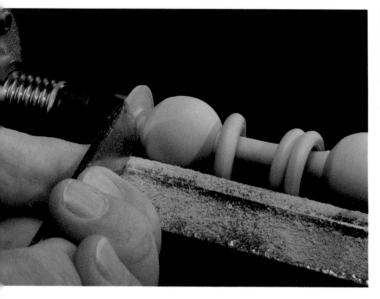

With the long point of the skew, reduce the stock on both ends, being careful not to cut the rattle free as we have not sanded it yet.

Starting with the tailstock first, because the headstock is doing the driving, use the long point of the skew, working back and forth, making light cuts with the tip of the skew until the piece comes free in your hand. Provide support with your fingers behind the ball.

Sand the rattle.

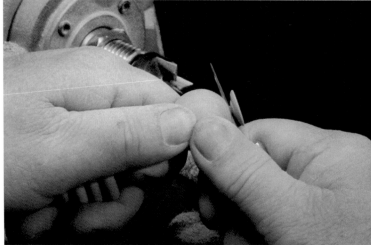

With a penknife cut off the ends and shape the very end if necessary.

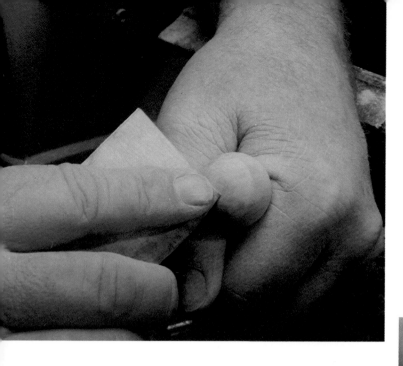

Sand the ends smooth.

Apply a liberal coat of mineral oil and allow to soak in before wiping off the excess.

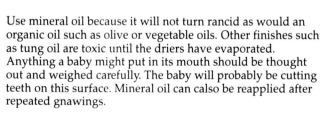

Use mineral oil because it will not turn rancid as would an organic oil such as olive or vegetable oils. Other finishes such as tung oil are toxic until the driers have evaporated. Anything a baby might put in its mouth should be thought out and weighed carefully. The baby will probably be cutting teeth on this surface. Mineral oil can calso be reapplied after repeated gnawings.

# Top

Every kid no matter what their age loves a top. There are a million different designs, but some work better than others. The one I like uses a dowel for the axis and a face grain piece for the body. We also must keep the center of gravity low to achieve a good spin. For the book, I am using cherry, a warm wood.

I have started with a 3/8" dowel approximately 4" long, and a block with a 2 1/2" diameter by 1" thick body. These measurements can all be modified. Use whatever is necessary to accommodate your design. Glue in the dowel so that 3/8" protrudes through the bottom of the body. This will be the tip of the top. The other end is obviously the handle.

Clean up the body with a gouge.

Insert the handle end of the dowel in the drill chuck. With the lathe running, pull the tailstock up and let it seek its center on the dowel before locking it down and applying pressure to the work piece.

Start shaping the angle to create the shape of a top.

Cut into the back side to help maintain a low center of gravity for the top.

When starting a cut with a gouge, stand the flutes in a vertical position. This allows the very tip to make an indentation for the bevel to ride. Once the indentation is made, rotate the gouge into the cutting position and proceed with the cut. If this is not done, the cutting edge has nothing for the bevel to ride on and will skate across your work piece.

Progress.

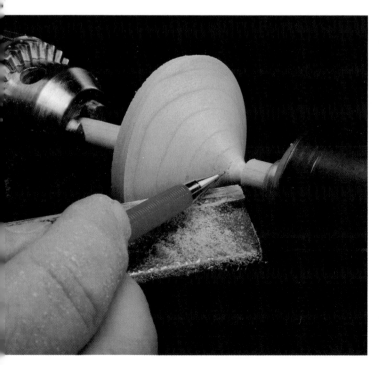

The bottom of the top is a problem area because the grain of the body is so thin at the junction with the dowel that it has a tendency to chip out, especially if it has not received a coating of glue when assembled. Take light cuts and when close to your desired shape, saturate the area with thin superglue. This will bond the wood to the dowel and help prevent chip out.

Using a gouge, shape a transition back to the dowel that will be pleasing to the eye.

Shape the handle.

Don't cut too far on the top of the handle because the handle top is going to have to drive the top when you sand and cut off the tip.

Progress.

50

Sand.

A fine cut.

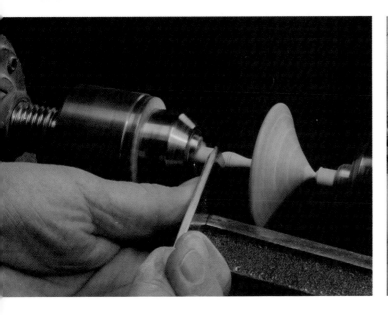

Add decorative lines with a long tip skew. Buff with the 400 grit sandpaper.

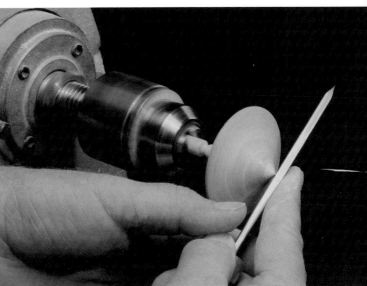

Sand the tip until it is very smooth because this is the surface the top spins on. Using the flat surface of a skew, with the lathe running, burnish the tip to work-harden the tip and give a longer spin.

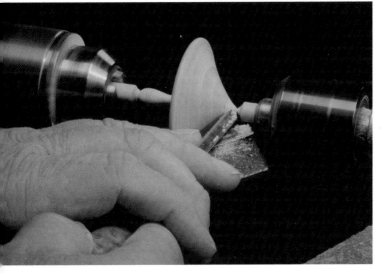

Cut the top free at the tail stock. Take light cuts remembering it is being held only by the handle.

With the long tip of the skew, take light cuts back and forth until you part off the handle, supporting the top with your hand.

Sand the end of the handle.

I chose to oil this top because I made a very human mistake. If I had not cut the top off the lathe before finishing, I would have used a French Polish and finished it on the lathe.

Tip top.

# Found Money Jar

From trash to treasure: My wife Cindy and I call this a found money jar--something we started to use to keep the money that we find on our walks, in parking lots, etc. This money we hoard and display in a jar as such.

I am using a piece of spalted birch for this project because it has an attractive figure and fine grain. Any wood or jar can be used. The possibilities are limited only by your imagination.

Choose a blank of wood larger than the lid that you will be covering. Attach the blank to the scrap block with double face tape. With the lathe running, bring up the tailstock, let it seek its center and apply enough pressure to hold the piece.

Clean up the face with the gouge.

True the block.

With a parting tool, cut a recess for the lid.

This has to be taken in stages so the lid can be fit inside the work piece.

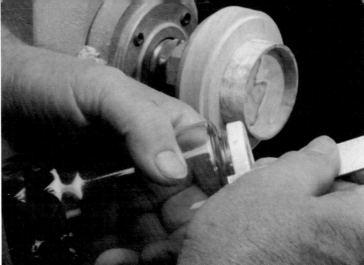

Determine the depth of the recess by measuring where you want the bottom of the lid in relation to the jar.

Keep fitting the lid to the work piece.

We have reached our desired depth. It is time to call it quits.

Switch to the skew. Use the long point to increase the diameter and keep the walls parallel. This cannot be done with the parting tool as the corners will foul out, creating a tapered recess.

Fits like a glove.

Sand the bottom edge and any exposed inner surface and give a coat of Deft™.

With a paper towel, wipe of the excess while the lathe is running. This coating of finish will help seal the wood and prevent cracking.

Remove the tape and use the scrap block to form a tenon to hold the lid.

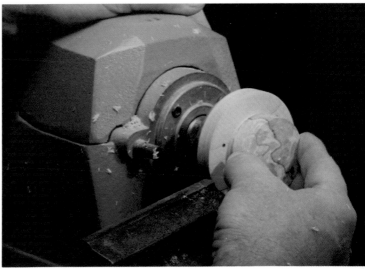

Proceed slowly and keep fitting to the cap.

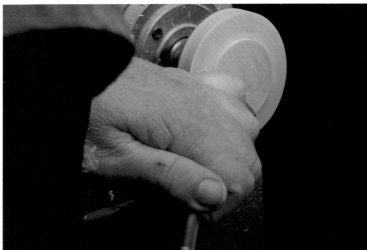

When making this tenon do not make the fit too tight as it will have a tendency to split the lid. The lid has to be just snug enough to drive, but not so tight that it will crack.

I've found it.

Clean up the sides a little bit.

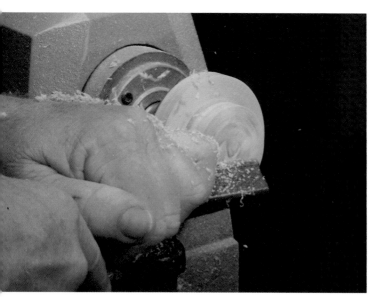

Shear scrape the top.

Sand starting with the 180 or your coarsest grit and work progressively through to the 400. After sanding the face, I pulled the piece slightly away so that I could sand all the way down to the bottom. Hold it lightly with your free hand so it does not come off the jam chuck.

Form a slight dome or concave surface as a flat surface is never as appealing. I chose a domed surface as you can see. Remember how deep your recess is so you don't break through.

Apply Deft™, and remove excess.

Use 0000 steel wool and oil.

Rough up the surface of the jar lid with sandpaper for better adhesion.

Use silicone adhesive on the inside of the wooden lid and set the jar lid in.

Finished. Beauty is in the eye of the beholder. To some a thick lid may be desirable, but I like the delicateness of this lid. Don't be afraid of failure in going too thin and breaking it. We don't have much money or time tied up in these, so if there is a faux pas, just try again. The extra effort of making a piece this thin can be very rewarding.

# The Gallery

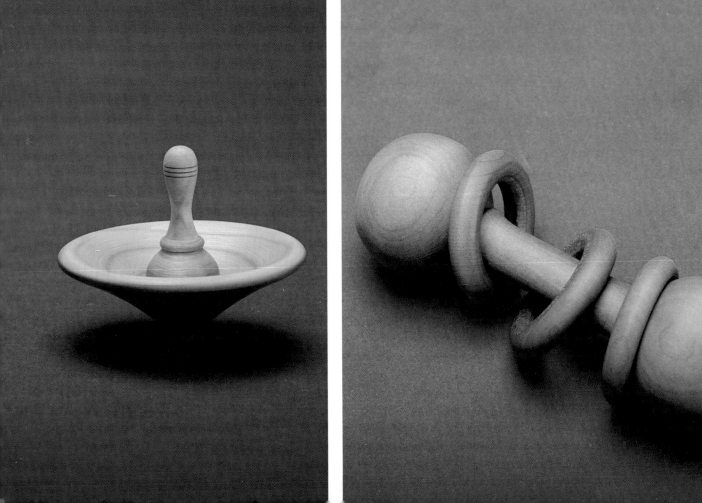